WHY WE NEED RAIN

CAN YOU FIND MY LOVE?

JAN MARQUART

Books currently available in the "Can You Find My Love?" Series

Seasons: Book1

Things To Do Outside: Book 2

Why We Need Rain: Book3

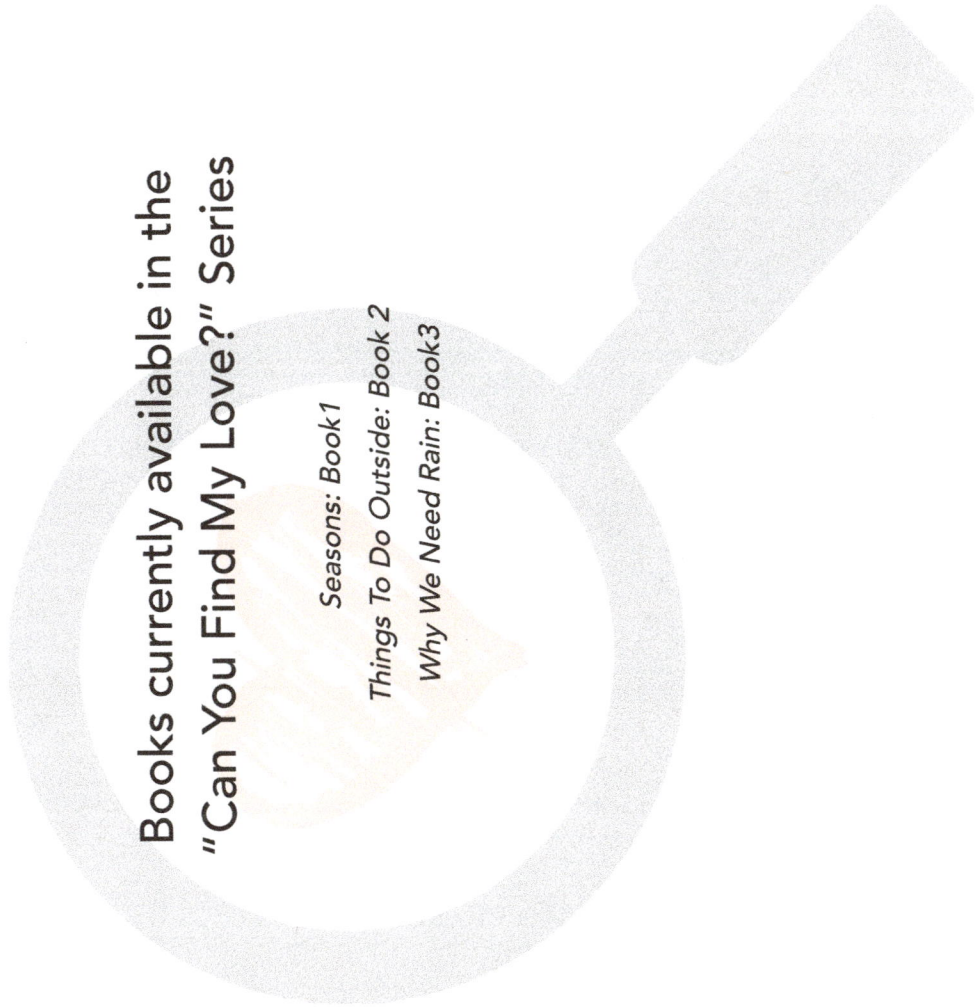

Other Books by Jan Marquart

FOR ADULTS

Write to Heal

The Mindful Writer, Still the Mind, Free the Pen

The Basket Weaver, a Novel

Kate's Way, a Novel

Echoes from the Womb, a Book for Daughters

Voices from the Land

The Breath of Dawn, a Journey of Everyday Blessings

How to Write From Your Heart (booklet)

How to Write Your Own Memoir (booklet)

A Manual on How to Deal With a Bully in the Workplace

Cracked Open, a Book of Poems

A Writer's Wisdom

To:

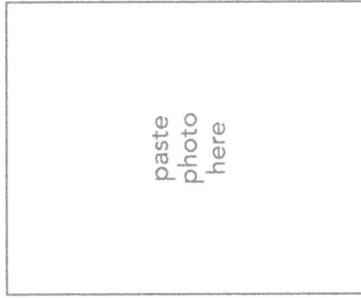

paste
photo
here

NAME

Thank you to all the parents, grandparents, teachers, doctors, daycare workers and others who have supported my efforts.

Also, my appreciation to Rich Carnahan, who continues to help me publish this book series, and his son, Aiden, a special young man who has greatly influenced my work.

CAN YOU FIND MY LOVE?

was inspired by two little angels:
Landon James and Evelyn Kirsten.

Their proud parents are my sweet nephew, David Maravel, and his beautiful wife, Shawn Maravel.

You have received this book because someone loves you.

Look closely—you will find love hidden in everyday things that you might normally take for granted.

This is what it looks like.

When you find the love I have placed for you, I hope that it warms your heart and lets you know how very special you are.

There would be NO water without rain.

Why We Need Rain

FOR DRINKING

We catch rain, clean it, and
pump it to faucets and fountains
for us to drink.

FOR ANIMALS

Rain provides drinking water for
animals that live in the wild
such as lions, bears, raccoons and deer.

FOR FARMS

Without rain, we would have
no crops to eat. Every fruit and vegetable
needs rain to grow.

TO KEEP COOL

Rain cools the ground on a hot day
which also keeps our air cool.

FOR RAINFORESTS

These hot, dense jungles get a lot of rainfall each year. Many plants from these forests are used to cure diseases.

TO MAKE MUD

Really wet, sticky dirt is called mud.
Without mud the ground would
crumble and crack.

FOR GARDENS

The plants, trees and flowers in our gardens
need rain to water their roots
and keep them healthy.

FOR THUNDERSTORMS

Thunderstorms are exciting.
They create loud booms, heavy rain
and lightning.

FOR RAINBOWS

When the sun's rays pass through
the mist after it rains,
beautiful stripes of color appear in the sky.

FOR PUDDLES

Puddles form when it rains a lot.
It is fun to put on your boots and
jump around in puddles.

FOR DROPLETS

Very small drops of rain are called droplets. Droplets on spider webs look like jewelry.

CAN YOU FIND MY LOVE?

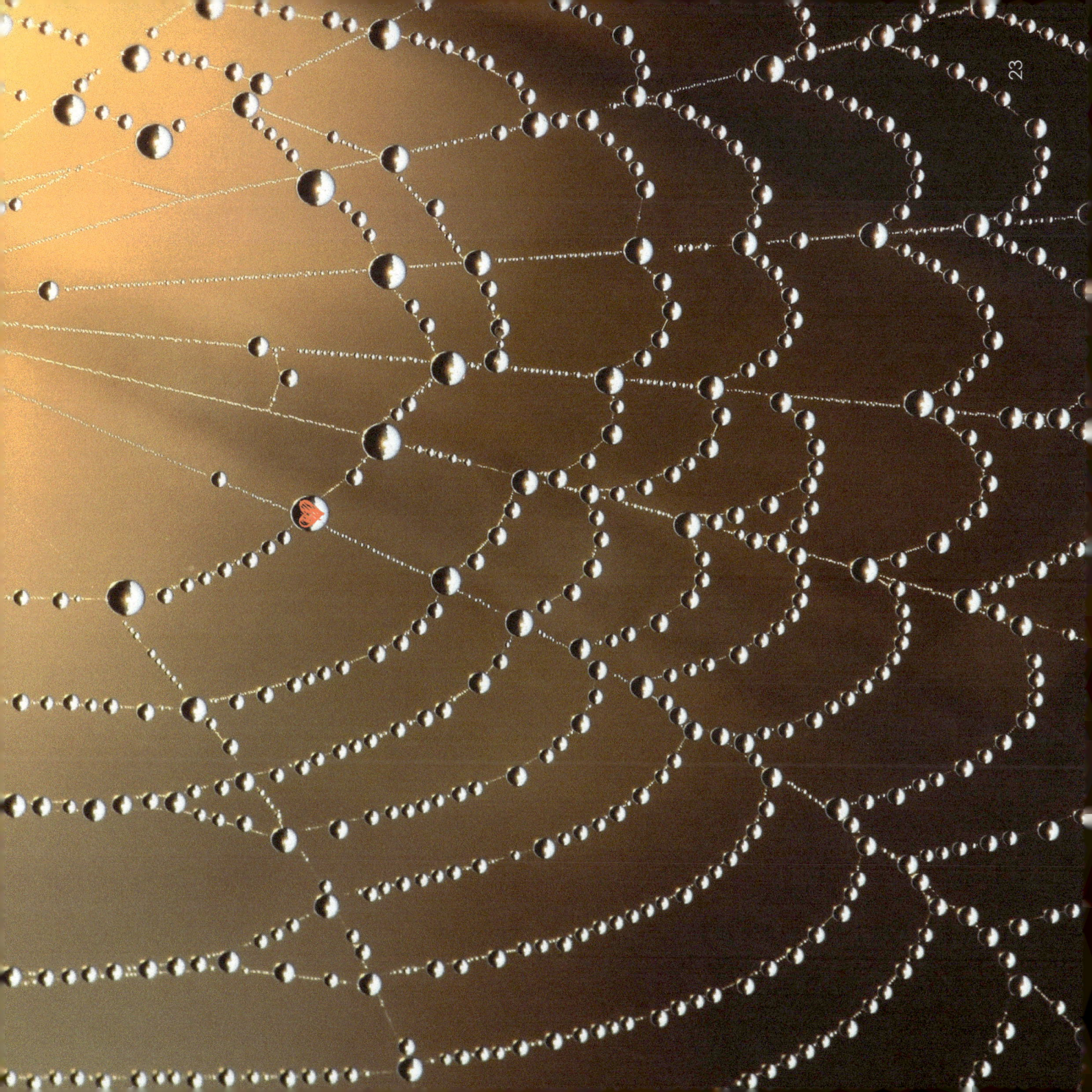

FOR BODIES OF WATER

Without rain, there would be no oceans,
rivers, lakes or ponds.
Everything would be dry.

CAN YOU FIND MY LOVE?

FOR FUN

There would be no swimming pools or waterparks without rain.
It's fun to play in the water!

FOR FISH

The water fish swim in comes from rain.
Fish can breathe underwater.

TO COOK

Many recipes use water.
We boil water to make
spaghetti and soups.

FOR WATERFALLS

When rain fills creeks and rivers,
the water that flows over the rocks and cliffs
is called a waterfall.

TO BATHE

We use water from rain to bathe and shower. It helps us stay clean and healthy.

FOR LAUNDRY

Rain gives us water to wash our clothes
when they get dirty.

FOR REFLECTIONS

Still water looks like a mirror.
Look into a puddle and you'll find your
reflection looking back.

FOR CLEANER AIR

As rain falls from the clouds,
it washes away chemicals, dust
and pollen in the air.

TO MAKE ICE

Ice is made from freezing water.
Cubes of ice are used
to keep our drinks cold.

CAN YOU FIND MY LOVE?

TO MAKE BUBBLES

You can make bubbles
by adding dish soap to a jar of water.

Did you look close enough
to find all my love?

Can you **DRAW** a few other things that need **RAIN**?

Can you **DRAW** a few other things that need **RAIN**?

Can you **DRAW** a few other things that need **RAIN**?

From:

NAME

paste
photo
here

About the Author

Jan Marquart is a psychotherapist and author. She has published 11 books for adults and has had articles, stories, poems and essays published in various newspapers, journals and magazines across the United States, Australia and Europe. She teaches writing for those over fifty and has taught a dozen writing workshops for Story Circle Network.

Jan has designed a 6-week writing course titled *Unveil the Wounded Self - Write to Heal* which focuses on healing PTSD and has also designed a 6-week writing course titled *The Provocation of Journal Writing* to encourage everyone to write their personal stories. She is currently on her 99th daily journal.

Jan can be contacted at JanMarquart.com, JanMarquartlcsw.wordpress.com and at her personal email address, jan@canyoufindmylove.com.

Her books can be purchased from all major online book retailers.